CLIMATE CHANGE - HEALTH CONNECTION

How climate change is affecting our health

NANDITA KHANDELWAL

ISBN: 9789355666024

FOR MY PARENTS

CONTENTS

FOREWORD

Climate change is a burning topic. If more and more people will propagate the term climate change, it will be helpful. Citizens of the country should know the climatic change would eventually harm various ecosystems flourishing on the Earth, in general, and humans in particular. The author has tried to explain the fundamentals, various components, and effects of climate change. Through this book, the author wants to shed some light on how climate change affects our health? And how can we protect ourselves from its effects? This book is most useful for school children and people who are not associated with various science subjects. Books that create awareness about climate change should be published as much as possible. This book is in the direction of saving the human race, our mother Earth.

- Dr. Dhirendra Kumar Pandey Professor and Head Department Geology Under the School of Earth, Biological and Environmental Science Central University of South Bihar SH 7, Gaya-Panchanpur Road

DOOR TO WHAT LIES AHEAD

Climate change is not new for planet Earth, but for a human being, it is. Currently, we are adapting, and we do not realise the repercussions of this change right now.

All the organisms can adapt, and if not, they will migrate. Currently, we are adapting to this environment. In the future, we will not be able to adapt and are left with one choice, to migrate from one place to another. Do you think it will be easier for billions of people to migrate?

Apart from this, currently, climate change is affecting our health. Climate change is the latent factor of various diseases. Every possible area has been affected by climate change. And we still believe that it is natural, and can not do anything about it. You can do this. We all can do it. In the Earth's history: human beings are the first organisms that can bring change.

You are the first generation who is going to suffer. I am writing this book to make you all aware of the root cause behind our sufferings and tell you the solutions which can bring changes in our day-to-day lives.

We all have to come together and think about innovative solutions to deal with this climate change, to reduce carbon emissions. #reducecarbonemissions.

THANK YOU

To my grandparent, Babu Lal Khandelwal who has always believed in my dreams and never doubted my capabilities, and my parents who supported me throughout my academic career.

To my Professors, Dr.Dhirendra Kumar Pandey and Dr.Aruna Pandey, and colleagues who constantly taught me the different aspects of Geology and Health.

To my coach, Sweta Samota who constantly guided me and provided sources to finish the book. The mysterious H, my editor for holding my hand through every meltdown.

To all my online and offline friends who helped me along the way including Pratish Kashyap, Tamanna Meena, Dan Watt, and Jainil Mehta.

Finally, to Aditi and Chandan, who don't give a hoot about any of this.

PART 1.

OUR EARTH: THE ONLY LIVING PLANET WE HAVE

1. A NEW UNDERSTANDING OF THE EARTH

Imagine, one morning, you woke up and found yourself at an unfamiliar place. You are standing straight, motionless, powerless, and your blood is running cold. You can't breathe, you find your mind at sixes and sevens, and your heart is heavy. You feel a void in your stomach and an urge to puke. You rub your eyes and still can't see anyone around you. There is no colour, everything is white. You struggle to find your friends and family. There is no sun, trees, or clouds. The breeze is chilled, and now you have started shivering, walking barefoot on that uneven surface. You put your hand in your pocket, and there is no cellphone. You are frightened. You let out a howl of anguish that not a single person has heard. On the other hand, you are having a second thought that everyone died. You try to recall the last night, but can't recollect any incident. In tandem tears of anger and helpless feeling fall from your eyes. You've got a frog in your throat. After all the smoke and mirrors, you get faint.

Your mother shouts and nudges you. *It was a dream.*

Relax, It was one of the dreams I've shared with you. We are still living on the same planet "THE Earth." It is with reverence for our planet that nourishes, sustains, and cares for us in a myriad ways that I'm writing this. I sincerely hope that I shall impart to my readers at least some of the intensities of the subject.

When I look at the Earth from the outside (satellite image), I feel all the problems are too small. We human beings are nothing but a speck of dust in this vast Universe. Whereas if you zoom in, you will find that we are surrounded by various problems and one of them is Climate change.

"I believe the simplest explanation is, there is no god. no one created the universe and no one directs our fate. This leads me to a profound realisation that there probably is no heaven and no afterlife either. We have this one life to appreciate the grand design of the Universe and for that, I am grateful", said by Stephan Hawkings.

In recent times, scientists are doing a lot of research on Mars. They are trying to find out the possibilities of life there. Scientists have also found some water on Mars, but that doesn't mean that we can shift to that planet immediately and start a new life there.

We all should feel blessed to be a part of a planet like the Earth which has all the key ingredients in the proper quantity to sustain life here. Let me share with you a piece of research that will give you a brief idea of how difficult it is to find a planet where life can exist.

Dr. Bredan Dyck, Assistant Professor at the University of British Columbia is exploring life on other planets, mainly Mars. According to him, if we are successful at finding the amount of iron present in a planet's mantle, then there is a great possibility to determine the thickness of the crust of the planet which in turn will help us to figure out the amount of liquid water and the atmosphere of that particular planet. *According to him, this is a feasible way to discover potential near planets where life exists.* All the small planets have one thing in common, that is they have the same portion of iron. However, to differentiate them, analysis of iron content in the mantle versus the core is mandatory. The thickness of the core is indirectly proportional to the thickness of the crust. For example, on the Earth, the thickness of the crust and the core are thinner and thicker, respectively. The thickness of the crust plays a critical role in exploring the atmosphere, water, and plate tectonics of that particular planet. These three aspects can tell whether there is life on a particular planet or not?

The above research is a reflection of the fact that it may take thousands of years for humans to find a planet where life can sustain itself. We should be grateful to be on a planet like Earth.

CLIMATE CHANGE IS REAL

When I was in college, there was a lot of discussion about geological time scale and climate changes. Climate Change is not a new concept for our planet. The climate has changed since human beings did not exist on the planet. Change is inevitable. There were periods of colder and warmer climates on Earth.

Geologic evidence suggests that one of these periods occurred 700 million years ago and there were many others. In this book, we are talking about the latest climate change on the Earth which can be a threat to humanity. In this era, climate change was first coined nearly 30 years ago, in 1989, in the medical literature.

Scientists predicted the mean temperature at the Earth's surface might rise from two degrees to five degrees in the next 100 years. Should we be alarmed that the ocean may rise by 1 metre in the next 50 to 100 years?

The answers are clear now. According to the latest studies, the global average temperature of the Earth will increase by at least 2 degrees Celsius by 2100. The sea level will rise to at least 1 metre by 2100. I understand many of you must be thinking 2 degrees Celsius only! How can it make a big difference in our lives? How will climate change impact humanity? Is this warming real?

Certainly, global warming is real!!

We are human beings, we are fallible. We react to what's right in front of our eyes, what we can taste, feel. The climate is very challenging. It is like gravity. It is always present but none of us pay attention to it. People are not curious to know the root cause of their sufferings.

"We are the first generation to feel the impacts of climate change and the last generation to do something about it." Said by Barack Obama

Climate change is not only about the future generation, but also, about the health of species currently living on this planet. Most of us are unaware of the impact it is having on our lives today and the lives of our children. Human beings have become the major force that is changing the course of our planet.

I know, we have a lot of "challenges," to face in this real-world besides climate change. We need to understand and come up with ways by which we can grow economically and financially and remain meaningful to the climate. Above all, we have one planet to live on. Let's make it a better place.

2. HOW IS EARTH EXPERIENCING CLIMATE CHANGE?

Earth's interior is divided into three major parts, crust, mantle, and core. In general terms, Earth consists of three important parts: the atmosphere (air and gasses), the hydrosphere (water), and the lithosphere (the Earth's crust and upper part of the mantle). The surface of Earth is 70% water and 30% land.

Earth has several gases in the atmosphere that trap heat, including oxygen, carbon dioxide, methane, and nitrogen oxide. These three major gases, along with other gases, retain heat and make the climate of the Earth what it is. This is the only suitable environment in which a life can exist. Likewise, the hydrosphere is composed of glaciers, ice cores, icebergs, oceans, lakes, groundwater, and water vapour.

In geology, there is a Law of Uniformitarianism, "The present is the key to the past" by James Hutton (Father of Modern Geology). It means by studying and observing the present environment, past events can be inferred. Scientists reconstructed past climates by following the same law. To understand the global climate, they have studied all the evidence and tried to solve this puzzle. Evidence has shown that the Earth has undergone slow and continuous climatic changes. These small and slow changes can have adverse effects on us for a longer time.

A multi-university research project known as CLIMAP (Climate: long-range investigation mapping and prediction) studied past climates. They collected evidence in every possible way. Let's discuss some of them.

Vertical Ice cores and sediments collected from Antarctica reveal that for hundreds of thousands of years carbon dioxide levels have paralleled temperature. It has two different measurements (carbon dioxide and temperature) that have matched for as long as we've had records. Ice is made up of oxygen and water so ice cores can provide temperature details by examining the oxygen-isotope ratio.

Evidence can also be found in the tree rings (dendrochronology); each ring represents a year. The changes in the thickness of the ring indicates temperatures that have taken place within a year.

Also, oceanic sediments from Greenland contain the remains of organisms that used to live near the surface. These sediments indicate the surface water temperature.

Other shreds of evidence that have been used to reconstruct past climates are:
Records of natural lake-bottom sediment
Soil deposits and sea sediment
Pollen in deep ice caves
Corals and stalactites in caves

Climatologists try to derive all the possible evidence to construct the past climates. When I was doing my Master's in Geology, I studied climatology and therefore became familiar with terms like shrinking ice sheets, glacial retreats, decreased snow covers, declining arctic sea ice, extreme events, ocean acidification, global temperature rise, and sea level rise.

In this chapter, I have discussed global temperature rise and sea level rise in detail as it is impacting our day-to-day life and health.

TEMPERATURE RISE

In this chapter, we will talk about the relationship between human body temperature and the Earth's temperature.

Factors contributing to the Earth's average overall temperature are:
1. Earth's orbit around the sun is more elliptical than circular.
2. How hot the sun's current temperature is over a period.
3. How reflective the Earth's surface is (white stuff, like ice caps, reflect light)

Recently, studies have shown that these natural factors are negligible. They are not the root cause of the temperature rise. The contribution of anthropogenic activities (that is human activities) is massive. The data has been collected from 1951 to 2010. Since the last climate change event, carbon dioxide from human activity is increasing more than 250 times faster than it did from natural sources.

Carbon dioxide is the most common greenhouse gas. It might not be the most potent, but the sheer amount of it has a large impact on our climate. Greenhouse gases play a vital role in the temperature rise of the Earth's atmosphere. There is no uniformity though. Since the late 19th century, on average, the entire surface of the Earth has warmed nearly 1 degree Celsius.

The expectation that adding greenhouse gases into the atmosphere would warm the planet is not a new science. In the late 19 century, Svante Arrhenius, a Nobel prize-winning Swedish scientist, wrote a paper in which he predicted that the outpouring of carbon dioxide from the ignition of coal would warm the planet to a great extent.

How much warmer our planet may get by the end of this century as compared to its start?

Most places will get warmer from 1 degree to 4 degrees Celsius or more. The difference between these two futures depends heavily on how much more greenhouse gas gets added to the atmosphere as well as how reflective the surface of the Earth is.

In India, people are generating carbon dioxide by burning fossil fuels: coal, natural gas, and oil. Coal is the largest contributor to carbon dioxide. Some amount of carbon dioxide is also generated via "direct emissions" - agriculture, mining, and forestry but these activities also produce most of the methane and nitrogen oxide that humans release into the atmosphere.

Carbon dioxide generated by human beings is only partially absorbed by plants and the oceans each year. However, the carbon dioxide that human beings are producing is building up over time and will take centuries to go away. It takes hundreds or thousands of years to leave the atmosphere.

Carbon dioxide concentrations are likely to reach 500 parts per million by the year 2100.

Choices about where our energy comes from, and how we manage the Earth's resources more broadly also carry immediate relevance to our health.

SEA LEVEL RISE

One day, I went to a cafe. I was sitting and reading an article on sea level rise.
A guy who was sitting next to me questioned if it was even true?

I replied, "Yes."

He said, "Isn't it a good thing? We will have more water in the oceans. Maybe then we can solve the water scarcity problem."

I smiled and asked him, "Do you drink ocean water? "Yes, sea level water is rising, but that water is salty.

He said, "We can use a purifier."

I replied, "For all the purposes, watering crops, bathing, etc?"

He smiled and said, "You are right."

When greenhouse gases trap heat, they not only raise the temperature of the air but also of water. More than 90% of the warming that greenhouse gases have added to our planet has gone into the oceans. As the water warms, it expands, and that makes the sea level rise. Sea level has already risen significantly around the world, by an average of roughly 90 mm (3.5 inches) since the early 1990s because of temperature rise. The rates of sea level rise vary from place to place.

Apart from temperature rise, sea level rise in the past century was primarily because of melting land ice. This land ice includes mountain glaciers and ice sheets, particularly in Greenland and Antarctica.

Thermal expansion of water has been the next biggest contributor to rising sea levels. In recent years, scientists have been debating the subject of the magnitude of sea level rise. Moreover, Ice melts on land have acted as a catalyst to increase the sea level rise by 2100. Earlier, we expected 1 metre of sea level rise but now with a better observation of ice melt, we can say it might even rise 2 metres. Again, some places may experience much more sea-level rise than others.

Ocean acidification

Water is present in all forms: groundwater, glaciers, ice, in the atmosphere but the major portion is in the ocean. The amount of carbon dioxide gets released into the atmosphere, about 30% of it gets straight away into the ocean. Oceans are a storehouse for carbon dioxide. As it gets absorbed, it produces carbonic acid, which is making the world's oceans more acidic.

How does the intensification of the water cycle affect rainfall where you live?

Changing levels of carbon dioxide show a shift in ocean circulation patterns. Such shifts enhance the negative effect on the water cycle. As we know that warmer air holds more water. As air temperature rises, the atmosphere holds more water. When that air with more moisture is ready to rain or snow it does so in greater amounts, heavier downpours are already prevalent around the world.

Climate models indicate precipitation events are likely to get even heavier if greenhouse gases emissions are not curtailed. As a general rule, climate change will make places that are getting wetter will get wetter, and places getting drier will get drier.

Scientifically, studies have shown that if it rains about 10 days, each of these days may receive on average closer to 0.5 inches of rain in 2100 rather than 0.3 inches as it receives today.

On the contrary, the warmer temperature causes more evaporation and drying land. Moreover, If greenhouse gases emissions are not reduced, climate models state that droughts severity is likely to substantially increase over this century.
In this chapter, we've covered the two major factors which are affecting our climate and our lives.

Do you observe these changes in your day-to-day life? If yes, tell me how? Invest some time here, where do you live? What are the common climate related issues you face? What can you do to solve it? If Yes, what is your government doing to solve this problem? Write your answers here.

Gentle Reminder:

Human beings are capable of reducing carbon emissions.

1) By following all the policies which are created by the government to protect our environment from climate impacts.

2) By taking small steps that will help you understand climate change better. Check out chapter "A Road towards a Better Future".

PART 2.

CLIMATE CHANGE AND ITS EFFECT ON HEALTH

3. HEAT, HUMIDITY, AND HEALTH

India is a tropical country, which makes it hotter than most countries. Tropical climates are where the noon sun is always high, day and night are nearly equal in length. The human body's internal temperature lies between 98.6 to 100 degrees Fahrenheit. If the body temperature exceeds this range then it is known as fever. The first thing a person experiences when body temperature rises is sweating. We all know that 104 degrees Fahrenheit is the limit for our bodies. After this, our internal body starts to break down and we require immediate medical attention.

Ultimate heat and humidity can cause diseases such as heat strokes in which the body overheats and figuratively gets cooked.

The heat exposure pyramid shows a range of health effects caused by heat exposure and the proportion of a population that may be affected. The wider the pyramid is at the level of the specified health effects, the more likely the health effect. However, death from heat exposure is rare.

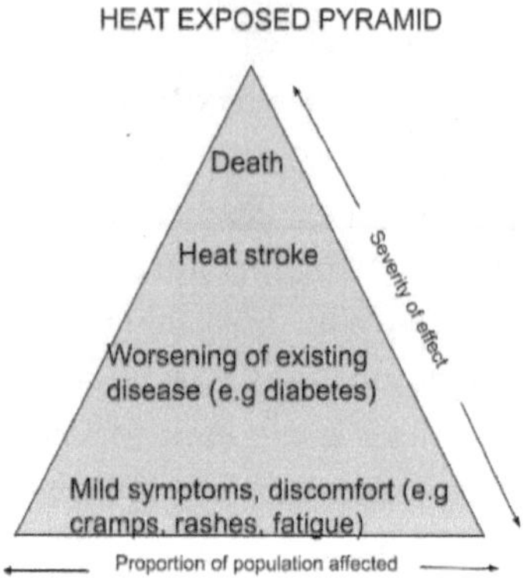

More often, exposure to heat makes the existing disease worse. Heat can promote dehydration which makes a wide range of diseases worse, including heart and lung problems such as asthma, chronic obstructive pulmonary disease, diabetes, and heart failure. This is a significant problem for obese and elderly, who have more difficulty in shedding excess heat. In India, around 12.4% of people are suffering heart related problems.

Heatstroke may also result in permanent damage to our brains and nerves and in worse cases even death. Heat can also slow down the working of our brain and thus can result in decreased productivity. The heart, brain, and lungs are all sensitive to heat stress and when they get overheated, they fail.

During heat waves, hospitals experience the rush for a variety of heart-related conditions such as heat exhaustion and heat stroke, heat cramps, dehydration and electrolyte disorders, cardiovascular disease, respiratory disorders, neurological conditions, and neutral illness. Furthermore, when childbearing women get exposed to extreme heat they may be at risk of the neural tube and other birth defects.

In recent years, severe heat waves around the world have led to the deaths of 10 to 1000 people, either from the effects of heat itself or because of air pollution that heat waves can generate.

Lethal heat waves are frequent across the world. Currently, new research has evaluated that 30% of the world's population is exposed to potentially destructive heat for 20 days per year. Without major depletion in the emission of greenhouse gases such as carbon dioxide, up to 3 in 4 people will face the threat of death from the heat by 2100. Unfortunately, by the end of the century, even if we reduce carbon emissions now, one in two people will suffer from ultraheat for at least 20 days, which can kill people.

I don't know why our community is not concerned about the danger?

Maybe because we don't pay a lot of attention to heat in the summer, because psychologically we expect it to be hot. Earlier, In India, the heat was not a huge problem but now heat extremes have a more intense impact on climate change. Annually, thousands of people are dying because of heat waves.

HUMIDITY

When the air temperature and amount of water vapour in the air are high, then the air becomes saturated with water vapour. This water vapour leaves moisture on our bodies, and it does not evaporate. The National Weather Service uses the heat index to determine serious weather-related health diseases. The heat index combines air temperature with relative humidity to calculate an apparent temperature. An apparent temperature has been divided into four categories based on the range of temperature. These apparent temperature categories determine the possible heat related problems in humans.

Category 1: 130 degrees Fahrenheit or higher will lead to heatstroke or sunstroke.

Category 2: 105 degrees - 130 degrees Fahrenheit causes sunstroke, heat cramps, heat exhaustion, or heatstroke is possible with exposure to sunlight and physical activity.

Category 3: 90 degrees - 105 degrees Fahrenheit give rise to sunstroke, heat cramps, heat exhaustion is possible with exposure to sunlight and physical activity.

Category 4: 80 degrees - 90 degrees Fahrenheit fatigue is possible with prolonged exposure to sunlight and physical activity.

As a Health Coach, I would recommend all of you take care of your water intake. Only hydration can beat this humidity and heat. Adult bodies require 3 to 4 litres of water per day. If having simple water is not possible, try to infuse your body with other sources like juices, lemon juice, and coconut water. Make health a priority and build a healthy lifestyle.
In rural areas, some people are still cooking their food by burning fossil fuels like dung, coal, and wood.

The burning of these fuels can generate particulate matter, which is harmful to people and our surroundings. Some people still believe that the food tastes better when cooked like this. Whereas in some places, people don't have enough resources to use a proper stove in India.

Household air pollution is a door for premature death, eye problems, heart problems, and respiratory diseases. In 2013, 9,20,000 people died because of household air pollution, and 5,90,000 people because of ambient air pollution.

This heat can have various other consequences. Heat can trigger wildfires, raise levels of ground ozone, allergies, etc. Let's discuss this in detail.

WILDFIRES

Climate change causes heat waves and more severe droughts. These heat waves create favourable conditions for wildfires. When forests burn they produce smoke that is composed of many toxic substances. Harmful substances in smoke from wildfires include acrolein, carbon monoxide, formaldehyde, benzene, polyaromatic hydrocarbons, and particulate matter.

Particulate matter refers to PM2.5 which is 2.5 microns in diameter or smaller. PM 10, which is considered less harmful, can still be dangerous.

PM2.5 is about 1/20th the width of human hair. Research on PM2.5 has demonstrated that when people breathe it in, it can be deadly, particularly by causing heart attacks and strokes. Worldwide, particulate matter exposures are responsible for millions of deaths each year. Particulate matter exposure has also been associated with preterm birth, lung cancer, and a host of other diseases.

The fourth highest-ranking risk factor for death in the world is exposure to air pollution. Particulate matters are derived from different sources apart from fires. For PM in outdoor air, the combustion of fossil fuels with coal and diesel fuel makes substantial contributions in most places. Indoor particulate matter air pollution is as dangerous as outdoor pollution however its sources are different.

According to research in 2015, contributions to PM 2.5 in India was 37% traffic, 4% industries, 16% domestic fuel, 22% from other human sources, and 21% from natural sources.

Domestic fuel comprises wood, coal, gas fuel used in various household chores like cooking and heating. Industrial emissions consist of oil combustion and coal burning in power plants.

Natural sources stand for dust aerosolized from fields or soils and sea salt. Sources of human origin refer to air particles formed from chemical reactions in the atmosphere, so-called secondary particle formation.

Besides this, forests are a natural habitat for wild animals. Wildfires lead to a loss of habitat of animals which causes animals to migrate from one place to another, this can be dangerous for human beings as well as for animals. The Ebola virus in Africa is an example of this.

GROUND LEVEL OZONE

Ozone is a layer in the stratosphere (atmosphere) that shields us from the sun's ultraviolet radiation. However, ozone at ground level (in the troposphere), where we can breathe it into bodies, is a different story. Ozone is a noxious substance with an unpleasant odour. The health effects of ozone on the human body are:

1 Headache

2 Burning (eyes, nose, throat)

3 Heart attack

4 Difficulty in breathing, cough, and chest pain.

Ozone is bad for humans and for plants too. It damages crops and retards tree growth.

Nitrogen + Volatile organic components + heat + sunlight forms OZONE. There are more ozone forms in the summer because there's more UV radiation hitting the Earth from the sun. Scientists have tried to forecast how a higher average temperature may influence ground level ozone in the decades to come, but the levels of ozone are hard to predict.

Ozone formation happens more readily at higher temperatures. Ozone pollution is expected to grow in many places around the world as the temperature warms across this century. Ozone mortality will be most pronounced in India and China, while worsening in the eastern US, much of Europe, and southern Africa.

Ozone harms plants. It damages leaf cells which in turn stops the growth of the plant. The production of fruit, vegetables, and wood stops if the ozone level is higher in the atmosphere. For a shorter period, plants can protect themselves from ozone by closing their stomata but it is temporary. If leaves close their stomata for a longer interval of time then the plant's growth will stop. Closing stomata for a longer period will create a hindrance for carbon dioxide, which is important for photosynthesis.

ASTHMA AND ALLERGIES

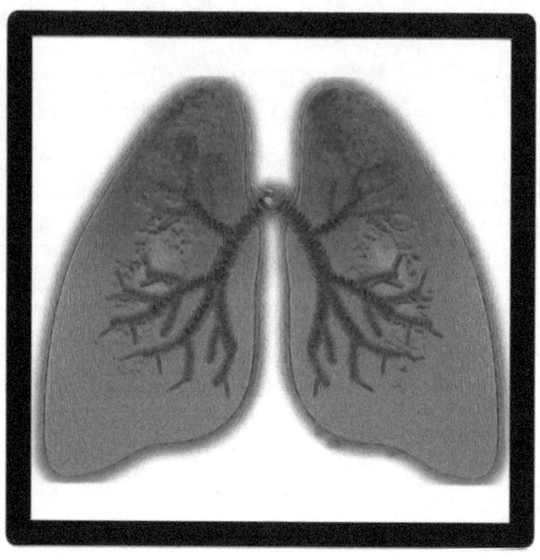

Warming temperatures have substantially lengthened the season during which plants that produce allergic pollen can survive. At the same time, increases in carbon dioxide in the air are contributing to greater pollen output. Pollen is a major contributor to seasonal allergies and can cause asthma attacks. Today the atmospheric carbon dioxide concentration is well above 440 ppm, which means the amount of pollen produced by plants may be twice that at the start of the 20th century.

Higher temperatures can lead to high weed growth which can adversely affect the production of various crops. Weed growth hurts our crops.
In more than 20 years of research, Lewis Ziska, a Ph.D. weed ecologist at the agriculture research service division of the U.S department of agriculture, discovered that weeds that flourish in the countryside are generally five to six feet tall.

Although in the city the same weed is ten to twelve feet tall. The cause is the higher percentage of carbon dioxide present in cities.

Pollen is the cause of allergies and asthma attacks. Carbon dioxide is good for plants, but excess is harming our crops. Weeds take all the nutrients from our main crops.

Heat is not only affecting human beings but also animals, forests, crops. We all are interrelated. This generation has to understand how to build the economy while taking care of wildlife, forests, crops, farmers, and public health.

If people still believe that everything is natural and they are not harming the climate then they are wrong. Climate change is here, but people are not ready to think about it. As we all can adapt, we are adapting every year.

What we don't understand is that there will be a point when organisms including humans will not be able to adapt to this climate anymore. We won't be able to adjust ourselves to this environment therefore we'll be compelled to migrate towards the poles. Let's move on to the next chapter that will convince you that climate change is affecting our health.

"Our attitude towards the environment has been so reckless that we are running out of good choices for the future."
Said by Mora of the University of Hawaii

4. THE EFFECT ON FOOD

In this chapter, I would like to tell you how nutrition and climate change are interrelated.

Nutrition is a broad spectrum. Currently, a lot of research and studies are going on. After all, everyone needs healthy food and a healthy lifestyle. Our body cannot function properly without nutritious food. Lack of nutritious food can lead to nutritional diseases which have an immense effect on the health of children, the elderly, and women (especially pregnant women). Each particular nutrient has its role in our body. An imbalance of any nutrient in our body can open the door for unwanted circumstances.

"Everything in excess is opposed to nature." said by Hippocrates

In nutrition, we have two broad categories. We have macronutrients that include protein, carbohydrates, and fats and micronutrients that include vitamins and minerals like iron, zinc, magnesium, vitamins A, C, etc.

Every nutrient has a remarkable role in our body. Excess or deficiency of nutrients may lead to many lifestyle diseases. Deforestation, coral bleaching, ocean acidification, floods, droughts, fish population, hunting, etc are the reasons for climate change which in turn impacts our food security and quality.

After working for 5 years in the fitness industry, I have observed that in modern days, the majority of the population is suffering from obesity. It is a major reason for various lifestyle diseases: high blood pressure, diabetes, joint pain, menstruation problems in women, heart attacks, strokes, elevated cholesterol levels, digestive issues, etc.

On the other hand, there is also a drastic increase in the number of the underweight population in India because of the lack of availability of the proper amount of nutrients in their diet. Being underweight is also a common reason for death. Moreover, it can result in lifelong suffering, especially in kids. Underweight kids become a soft spot for infectious diseases. A lack of nutrients impacts a child's overall growth and slows down brain development.

Do you have any medical conditions? Do you know anyone in your family who is suffering from lifestyle diseases? Write your answers here.

Every nutrient plays a vital role in our body. Human beings get affected by any nutrient deficiency in their body. One of the typical nutrient-deficiency diseases is anaemia. Conventionally, it is found in women. Women who are expecting should oversee the amount of iron intake present in their diet. 20% of maternal deaths happen because of anaemia alone. Moreover, if young children are iron deficient it will become difficult for them to balance it in the later stages of their life.

"Tiny things can be transformative" said by BJ Fogg

A lack of one nutrient can be so transformative that it may have a negative impact. Iron is crucial for every stage of life. Likewise, every mineral has a significant role in our bodies at every stage of life.

800 million women worldwide are suffering from anaemia, because of which they have painful periods throughout their lives. Furthermore, it may affect them during their pregnancy as well, also their kids. It is like a never-ending cycle. Women suffering from anaemia are bound to pills for iron, vitamin C, and other medicines as prescribed by their gynaecologists.

For example, one of my clients, a 24-year-old woman, used to work out every day. She was taking care of all her four meals: breakfast, lunch, snack, and dinner (all vegetarian). But still, she used to have painful periods. She used to feel lethargic and exhausted. It was not something new for her. She had this pain for a very long time. I advised her to visit a gynaecologist. Afterwards, she found out that she was anaemic.

Even after having all healthy, home-cooked food all day and night, she was disheartened with the results. Her body was deficient in iron content.

Currently, about 200 million people are deficient in micronutrients, like zinc, iron, magnesium, vitamin A, etc. They may struggle with sleep, lack of energy and strength, reduced work productivity and increased risks for other diseases and health conditions.

Now you must be thinking about how all these things are related to climate change?

We often hear our adults saying that things were better before, but now things are not pure. Do you know why? What has changed in this interval?

CROPS AND CARBON

Carbon emissions and greenhouse gases do affect our crops as well. It affects their nutritional value.

Dr. Sam Myers, director of the planetary health alliances at Harvard University and a principal research scientist at the Harvard TH chan school of public health, has performed a beautiful experiment to prove that a crop's nutritional value gets affected by greenhouse gas emissions. He and his team worked with a very large group of agronomists. These agronomists were from different parts of the world, to understand different climatic conditions and different scenarios. Australia, Japan, and two states of the US participated in this experiment. This experiment was known as FACE; a free air carbon dioxide enrichment experiment.

In this experiment, they had grown a certain amount of crops which were fielded by a ring of carbon dioxide emitting jets. These jets were continuously producing carbon dioxide in the field. Plants that were surrounded by carbon emitting jets versus plants in a normal environment had the same soil, pesticides, weather, and pathogens. The only difference was carbon emitting jets.

After comparing the plants of both environments, they concluded that crops looked similar in appearance. However, plants having excess carbon dioxide exposure had lower nutritional value. They were lacking in iron, zinc, and protein.

Especially crops like wheat, barley, and rice; were showing a drastic change in iron, zinc, and protein. On the other hand, crops like legumes, soybean, and peas have shown lesser amounts of iron and zinc but not protein. Crops which we rarely eat like maize and sorghum have shown minute changes.

Sam Myers, says that most of the models suggest that crops like wheat and rice production will fall in the tropics. In upcoming years, carbon dioxide content will be 550 parts per million. By then, our nutritional value of food may reduce further. In India, people eat a lot of carbohydrate-rich food. It will have a major impact because no matter if we are living in the southern or northern part of India, rice, and chapati, is the prime part of the meals.

If those crops' nutritional values drop in terms of iron, zinc, and protein. It will be difficult for our vulnerable population to get rid of its consequences. Children and pregnant women are the prominently vulnerable populations in India.

They will suffer the most because of the drop in nutritional value of crops, as nutrient intake is mandatory for every human being to build up and reconstruct their body cells.

When I was studying at University, nobody discussed the effects of climate change on health. After reading this research by Sam Myers, I understood the reason behind their sufferings and the relation between climate change and health. The so-called nutritious food we are eating today might not be fulfilling the needs of our bodies because of its decreased nutritional value.

OTHER CHALLENGES AND CROPS

There are problems like water scarcity, land degradation, and global warming. All the climatic factors are going to affect our food production and consequently affect our health. It is a continuous cycle that nobody will ever benefit from.

GROUND LEVEL OZONE

Ground Level ozone formation and temperature rise will affect the quality and quantity of crops. Ozone toxicity has already damaged crops. As mentioned in the previous chapter.

SEA LEVEL RISE

Temperature rise and greenhouse gases emissions will increase the sea level which might cause floods. A change in the rainfall pattern is expected, as all the climatic models have predicted . The temperature rise in arid areas will bring about severe droughts.

Droughts will enhance water scarcity, groundwater levels will go down, and irrigation for crops will become difficult.

SALINIZATION OF COASTAL WATER

Population growth is fastest near the coastal areas. A temperature rise will raise the sea level which in turn makes the water saltier. Salinization of coastal groundwater creates difficulty for crops as they need freshwater for irrigation. No matter where you live, coastal or arid, the temperature rise will affect every one of us in some way or another.

POLLINATORS, PATHOGENS, AND PESTS

Insufficient intake of fruits, vegetables, nuts, and seeds in the diet is accountable for non-communicable diseases like heart attack, stroke, diabetes, and some cancers. The production of fruits, vegetables, nuts, and seeds is dependent on pollinators. The temperature rise has shown some repercussions on pests, pollinators, and pathogens as well.

SEAFOOD

Till now we were talking about the crops we eat, let's have a look over the seafood as well.

The most common seafood for Indians is fish. Seafood provides lots of nutrients that are required by the body like protein, healthier fats, and vitamin D. In coastal regions, a lot of protein intake in the diet is from seafood, especially fish.

93.4% of global warming is occurring in the oceans. We generally say global warming is about a warmer atmosphere actually, this heat is trapped in the oceans. When things become unpleasant for animals they migrate. Therefore, aquatic animals are also migrating toward the poles. The tropic regions are becoming too hot for aquatic animals to handle. By the end of this decade, certainly by the end of this century, people might have to deal with food insecurities. Nutrient deficiency which we are all experiencing due to the rise in carbon emissions affects the quality of crops.

Furthermore, due to the rise in sea temperature animals might migrate, which leads to food scarcity.

We cannot have supplements, protein powders, and pills for everything. During the COVID-19 pandemic people became aware of the importance of immunity and nutrients and started taking multivitamins, vitamin C, and vitamin D. Deficiency of micronutrients might lead to diseases. If eating supplements helped everyone then the government would have distributed them. But the fact is our body needs natural and nutritious food. The right amount of calories, nutrients, and micronutrients are needed for a healthy body.

Ground level ozone formation, temperature rise, and carbon dioxide levels are affecting the water we drink, the air we breathe, and the food we eat. If all the basic needs of human beings are getting affected by climate change then don't you think that it's high time to take some steps?

We all have to die someday. Dying because of any disease or disability which has occurred due to inappropriate food is sad. Earlier people struggled for food, therefore practised hunting, and invented agriculture, fishing and cooking. Today, we are in a world of advanced science and technology which can help us substitute our current habits with new ones to have positive outcomes. We need to work towards creating a food system that can provide nutritious food to people. Moreover, it should be accessible, affordable, and NOURISHING at the same time to everyone.

Gentle Reminder

Something which you all can do for yourselves is have healthy and nutritious food. Moreover, try not wasting food.

5. IS COVID-19 RELATED TO CLIMATE CHANGE?

Coronavirus was a new disease that had affected the entire world. It had challenged our immunities as well as our country's immunity. Not only was the health sector shaken up but also the economic, social, and other sectors. Research in every sector is directly or indirectly related to coronavirus these days. Let me explain how this coronavirus and climate change are related.

Coronavirus and climate change have the same heart, which is air pollution. As we all know, coronavirus attacks our lungs and immune system. Air pollution can act as a trigger for coronavirus, especially for people who have problems related to breathing and are addicted to smoking because they have sensitive lungs already.

In India, a massive number of COVID-19 cases were in the most crowded places and the places with the poorest air quality.

Metropolitan cities have a large population and poor air quality, hence we had to become more cautious. Our vulnerable population, like the elderly, asthmatic patients, and the poor needed a lot of attention and care.

Climate change is that latent factor that is leading to a lot of health problems and natural disasters. In the next ten years, we don't know how coronavirus or maybe any new virus or climate change may affect all of us.

Extreme climatic conditions like hurricanes, floods, tornadoes, and wildfires triggered by climate change forces the inhabitants to evacuate the habitat as soon as possible. When people have to migrate from their native place to any other place it becomes the best place for the transmission of infectious diseases like coronavirus, tuberculosis etc.

Not only people but also our entire ecosystem, economies and life quality suffer.

We must show some concern for our environment. How can we create a clean and pollution-free environment? Yes, washing hands and maintaining social distance has helped us. However, in the long run, climate change can have a massive effect on us.

If the air we breathe, the water we drink, the food we eat, and the shelter we have will not be safe and clean. Then where are we heading? Today it is coronavirus but soon it could be any other virus. We should ponder about our posterity and our health, which might suffer because of this climate change. We cannot deal with hurricanes, earthquakes, wildfires, tornadoes, etc, along with coronavirus.

OTHER INFECTIOUS DISEASES

Well, as far as we understand, there is no direct link between coronavirus and climate change. However, temperature rise has an impact on organisms of the Earth. Therefore, when the temperature rises, animals start migrating toward cooler regions. The spread of malaria and dengue increases from rising temperature, moisture in the air, and rainfall. The presence of pathogens is directly proportional to temperature and rainfall patterns. You will be astonished by hearing the news that the World health organisation has declared on 29 July 2021 that China is a malaria-free country after struggling for 70 years. India is still struggling with malaria, coronavirus, and various other infectious diseases.

Ebola virus is another type of virus that has spread across Africa. The main cause of this virus is deforestation, which is a continuous concern for humanity. It can be hazardous to humanity. Deforestation has led to a loss of habitat for animals causing wild animals to migrate. Wild animals in the city have become the root cause of the spread of various infectious diseases that are hazardous to humanity.

Infectious diseases have risen over the last decade, especially in cities because of the increasing population, temperature rise, rainfall, and deforestation. Animals found shelter where human beings are already living. It becomes easier for pathogens, mosquitoes, and different organisms to spread effortlessly.

If the Earth suffers, we are the ones who will suffer too. Climate change and health are not two different entities. 65 million years ago, dinosaurs went extinct. Almost half of the organisms went extinct because of various natural calamities. But we are human beings, who have created a whole other level of Earth, and we can bring change if we are mindful. That's what we call mindfulness in the health industry when a person is present in the moment and aware of his surroundings and their actions. Mindful activities make you well aware of your conscious mind and stress free in general.

Most of the population thinks that climate change is not in their hands and they are not harming the environment. Others believe the government should take actions, change the system because individuals cannot do anything about climate change. However, the truth is, we are all responsible for it. If every one of us takes responsibility for our actions then we can create a safer environment for our future generations.

PART 3.

HOW CAN WE HELP OURSELVES?

6. A ROAD TOWARDS A BETTER FUTURE

By now, you must have understood that solving the climate change problem doesn't mean helping only the Earth. It also means helping ourselves too. It means taking care of public health and future cohorts. We should take accountability for our actions and become responsible citizens of our country. We all are in charge of this change and can change it for the better. No matter where you live, the climate is also a global issue like coronavirus.

If you are a student, you should think about the problems you may face after ten years. If you are an adult, you should think about the challenges your kids and parents will face. If you are the oldest person in the family, then think about your grandchildren.

In this chapter, I want to share some solutions which we all can perform for ourselves and our loved ones. The four main areas which need the most attention and can give the best results are food waste management, transportation, buildings, and electricity. Yes, I understand that the government plays a vital role in any country, and they are doing their best. Furthermore, we need to become responsible citizens. If you are reading this book that means you are an educated person, a privileged one also. In India, a large population cannot read and write. Many vulnerable populations are poor and are disabled. We have to take care of them as well; so that we can create a better tomorrow.

As we have discussed a lot about temperature rise, let's start with that only. Cities are warmer than other areas because of the construction of cities, the population of the city, and pollution of the cities.

URBAN GREENING

Cities are famous for their roads, at least in India. Roads are always grey, which absorbs a lot of heat from the sun. Furthermore, buildings usually have dark roofs, which act as a good absorber of heat. Cities become warmer during the daytime, and it stays warm even at night because of pollution created by automobiles. In scientific terms, urbanisation leads to artificially raised average temperatures as cities grow. This is known as the Urban heat island effect.

This heat has already exceeded the amount of warming in recent decades that may occur due to greenhouse gas emissions. What we can do individually is to make our roofs green. Green roofs not only help in reducing the temperature of the city but maintain your building's temperature. Buildings will be cooler during the summer, and not only this, during winters, it will act as an insulator (they have less heat loss). Apart from this, it will reduce your energy consumption, and you have to pay fewer electricity bills and breathe more fresh air. This concept is known as Urban Greening.

I have mentioned the benefits of urban greening during the summer and winter seasons. What about the rainy season? In India, the rainy season occurs over a major portion of the year. During the rainy season, green roofs prevent water runoff into storm sewers. India faces sewer overflows, which in turn set the stage for all water-borne diseases.

Green roof buildings will have plants that absorb water and prevent water runoff, water-borne and infectious diseases. If you can't make your roof green by planting, try to paint it white. White colour doesn't absorb heat, and it will show benefits to the building's occupants in terms of heating and cooling.

TRANSPORTATION

In India, air pollution has been caused by burning fossil fuels like coal, diesel, petrol, and natural gas. If, we need to reduce the number of deaths happening because of air pollution. We need to minimise the usage of these fossil fuels.

Europe has already started doing it. People either walk or ride bicycles no matter how long the distance they have to cover. The Paris agreement is a legal document signed by various countries which focuses on climate change, of which India is also signatory. This agreement aims to reduce greenhouse gas emissions. The manoeuvre of renewable energy is not happening at the same pace as it was introduced in the Paris agreement. The change is happening, slowly and steadily.

Electric vehicles are entering the market but not flourishing yet. Till then we should amplify the usage of bicycles, metros, and public transport will help in the reduction of carbon emissions.

Apart from reducing carbon emissions, walking and riding bicycles are the best forms of exercise, which you can do for yourself. It will keep your respiratory system healthy. Walking alone has more than twenty benefits for your health. During the pandemic, physical exercise became an integral part of our lives. If you do exercise keeping your environment in mind, it will benefit you and your future generations. Instead of going to the office in your car/bike, you can start riding bicycles. In India, during the summer, it is not possible to ride a bicycle for long distances but what about the winters? It will not only save your money from buying expensive petrol or diesel for your car/bikes but keep your health on point.

Indeed, it is more feasible for the people who have their workplaces nearby. People who travel a lot of kilometres can use public transport like the metro or uber share. Over and above, this will save you money too. If you are avoiding people to maintain social distance and to save time, then take your car and ask your colleagues to join you. So that 2-3 people are at least travelling in the vehicle and reducing carbon emissions. This concept is becoming more popular nowadays and commonly known as carpooling.

When it comes to aviation emissions, present day aviation fuel needs to be replaced with carbon-neutral or low carbon liquid fuel. So far no alternative fuels meet the expectations of the aviation industry for efficient cost, availability, and performance. Scientists are still exploring the possible ways to reduce carbon emissions.

It is important for cargo ships to switch to low-sulphur fuels, which will reduce their contributions to air pollution but may not decrease greenhouse gas emissions. There is still room for improvement through ongoing research.

BUILDING

India is a developing country. There is a lot of construction going on, whether it is about constructing more buildings for public or personal use. The simple reason for an increase in construction is the increase in population. The requirement of more buildings for growing business, industries, and companies will keep increasing in the upcoming years. 40% of carbon emissions in the world are caused by the buildings using electricity produced by fossil fuels. If the population grows without climate change awareness then more carbon emissions will increase leading to various hazardous events.

People believe that metropolitan cities hold more opportunities consequently, migration from rural places to urban cities is happening so rapidly that in upcoming years, we may face a shortage of space. Thus, the Green building movement can mitigate the impact of human activities on the environment.

A Green building means a building that is energy efficient and saves natural resources like water. It creates a healthier environment for workers compared to conventional buildings. It also saves on water and electricity bills. Studies have shown that a green building's environment raises the cognitive performance of workers. The study, published on October 26, 2015, has shown that people who work in better air quality have higher cognitive function performance. Getting fresh air and sunlight early in the morning uplifts your mood. Unfortunately, these days we spend 90% of our day inside the office. During coronavirus, it was 100% indoors. Research has shown the negative impact that staying indoors had on us yet allowed nature to thrive.

The Green building movement is a well-known and results-proven movement that originated in the US Green Building Council in 1993. This movement is the best way to bring down carbon emissions which emit out of buildings, providing people with healthy, comfortable and efficient space. If we do the calculations for this green building movement, it will save 1% of the cost of energy. Moreover, 90% of the cost of green buildings can enhance health, air quality, and work productivity which can act as a great payback.

John Mandyck, Chief Sustainability Officer of the United Technologies Company says that the Green building movement is also a great business-building movement as you can increase occupancy rates even with higher rent rates. Green building is environmentally responsible and resource efficient which helps us economically as well.

Decisions that we make today for our country will show results in the upcoming decades. In India, we adopted the green building movement in 2001, embraced by the confederation of Indian Industry (CII).

How many Green buildings do you know are present in India? Write your answer here.

ELECTRICITY

The most efficient method to prevent carbon emissions is to replace fossil fuels with renewable sources of electricity. In India, electricity is still an issue in many rural places. Scientists aim to provide cheap and clean energy to empower impoverished people.

Austin Blackmon, Chief of Energy, Environment, and Open Space from the city of Boston sees an opportunity to bring people out of poverty. He said, we have to provide them access to electricity, but make sure it is clean electricity.

Fossil fuels contribute to hundreds of thousands of deaths in India each year. The cheapest and best sources of electricity are solar and wind but even they come with drawbacks. Michael Aziz, professor of Material and Energy Technologies at Harvard University, said that if the wind doesn't blow or the sun does not shine we still want electricity to watch television, cook, etc. In that case, you have to burn fossil fuel or use another renewable source of electricity. Therefore, batteries are the safest and most cost-effective way to solve this problem as we have batteries in our cars and phones. Batteries are not dependent on wind, sun, or rain. However, we have to recharge batteries so we can use them throughout the day.

Michael Aziz and his team are working on flow batteries. Flow batteries are best suited for stationary applications. The challenge they are facing is to produce batteries that are cheap, safe, and large-scaled. Our phones and car batteries are small, and they are not enough for large installations like homes, offices, and industries. We need batteries which can get stored in our basement or fit in smaller areas. These batteries should be able to run these large installations for a substantial time.

If we adopt small habits to reduce carbon emissions, lots of changes and outcomes can be seen in the upcoming years. There are several ways to reduce carbon emissions, for example: try to spend some time with your family and make your roof greener and use electricity wisely not only at home but at offices and public places too.

FOOD

For your better understanding, I have divided food into two categories; food production and food consumption.

Food production

Emissions of greenhouse gases while producing food are recurring. Using fossil fuels for the growth of food is conventional. Various farm equipment, including tractors and harvesters which run on liquid petroleum. However, there is not a lot we can do about this.

The production of pesticides and herbicides includes liquid petroleum gases and natural gases as crucial ingredients. Likewise, in the making of fertilisers, we use natural gas. It is normal to use chemicals, like pesticides, herbicides, etc in farming. On the other hand while practising organic farming, green manures, compost, and biological pest control we are reducing energy consumption and increasing the nutritional value of food. Therefore, organic farming should be our approach.

There are also various techniques like crop rotation and companion planting to enhance the nutrient value of soil. Besides this, a lot of farmers are also facing crop wastage. Crop wastage is happening during the change of seasons and sometimes due to uneven distribution of crops. Thus, cold storage facilities can be a better solution. It will not only keep their crops fresh but also enhance the life span.

Food consumption

The five components of our balanced meals are carbohydrates, fats, protein, vitamins, and minerals. We will discuss how one of the components is affected by our environment. Protein is essential for our body, but what form you are consuming can turn the tables on our environment. I am not asking anyone to shift their diets from red meat to lean meat, but I want you to be aware of the pros and cons of both. Studies have shown that red meat produces more carbon per unit of protein. Red meat also leads to a lot of other health-related issues. People who consume more red meat have a higher probability of getting heart disease, cancer, and premature death.

LESS RED MEAT
LESS HEAT

Organic food production does not require herbicides, pesticides, or fertilisers. They promote a reduction in carbon emissions and enhance our health too. However, nutritional values of crops are deteriorating by carbon emissions and red meat has its own repercussions. The question that arises is how to derive healthier sources of protein for individuals and the environment?

As a coach, I recommend lean meat: fish, legumes, nuts, poultry, and whole grains. Of course, the portion of protein in your diet plays a vital role. To understand the amount of protein required by your body you need to consult a dietitian or a nutritionist.

Did you know that the food you waste can act as an emitter of greenhouse gases? The production of food in the world is a lot more than we need to consume. There are approximately 8.1 billion people on Earth. The UN said that the food we discard can produce an ample carbon emission, approximately 4.4 billion metric tons of carbon. Wastage of food is the third-largest reason for emitting greenhouse gases. It is a silent attacker for climate change.

We all need to realise the goal of not wasting food. It is not only about your money. Food is still not affordable nor accessible to some people. We need to work on our food distribution policies. Food going to a landfill will not benefit any of us. Wastage of food has an immense impact on finances, nutrition, water, and climate. However, multiple steps can be taken to solve these problems.

By following the tiny habit of not wasting food at a party, or at home we can solve this problem. Whether that food is for free or we paid for it. If everyone takes responsibility to prevent the wastage of food, carbon emissions can be reduced.

"Charity begins at home, and Justice begins next door"
said by Charles Dickens

For Example, in India, people at a wedding, take a lot of food on their plates and throw any leftovers in a dustbin. People think that they haven't paid for this. Why? I agree you haven't paid for this but why waste it? If you are reading this book then please educate people around you on how their small actions can save our climate. It is not only about climate, but food can be consumed by someone else. After the celebration is over, we can distribute that food to poor people. Food is precious not only at weddings, but wherever you go, wherever you eat you can distribute it.

For example, there were three friends. They went to a new restaurant to try some Italian food. They asked the waiter to bring the signature food of the restaurant. The smell of that food was savoury, but the moment they had it, it tasted unpalatable. They paid the entire bill and left the remaining food on the table.

Then they went to their old restaurant and had North Indian food.

In your opinion, what can be an alternative here?

Yes, there is a choice. Instead of leaving their food, they could have asked the waiter to pack it. They can distribute it to someone who is in need. A lot of individuals do not eat even a single meal per day. People beg for food and money. Alternatively, they can bring the food home and share it with their family members. It is practicable that any other family member might like it. Leaving that food right there will not help you or any of us. It will go inside the dustbins and then landfills in turn, which will create *greenhouse gases*.

As responsible citizens, we all can participate in this for our environment. The choice is ours. Nobody is watching us; nobody is going to scold you or punish you.

"You reap what you sow," said by King James Version, in 1611

Whatever we sow today, our kids are going to get the fruit of it. If you are already a kid, then without change you might suffer in your adulthood or old age. Think twice. Be wise. Be a responsible person for yourself and your environment.

7. HOW CAN DIFFERENT ORGANISATIONS SUPPORT?

This entire book has suggested the choices we have. Yes, we have different solutions. We can choose to live the same life or start thinking about the future. Our choices define who we are, and what we stand for. This time our options will have an impact on our health and future as well. Our actions can still decrease greenhouse gas emissions. As coronavirus has challenged our immunity, climate change will also do the same in the future. None of us are capable of resisting it.

We are all in this together, and we need to acknowledge the importance of this change. During coronavirus, many organisations came together to help and support their countries. Likewise, we need to do this together including schools, banks, NGOs, private organisations, and the government. We can participate in reducing carbon emissions and we should.

ROLE OF ACADEMICS

Schools are the primary source of information for kids and therefore should educate our youth about the impacts of climate change on our health. As Barack Obama said, "This is the first generation to feel the impacts of climate change and the last generation to do something about it."

Kids need to understand every possible aspect of climate change. Communication about this issue in each house is significant. Creating awareness about the surroundings can be impactful.

India is a developing country that requires a lot of doctors, engineers, and scientists. But schools are the fundamental place where every engineer, doctor, and scientist has been educated. Engineers should be aware of what kind of machines and devices are needed to save our planet. Doctors need to understand how climate change can affect our health and what we should do to avoid harmful effects.

Scientists are already discovering various ways to reduce carbon emissions. On the other hand, not everyone wants to become a doctor, engineer, or scientist but everyone should be well versed about climate change.

Schools can provide research labs and demonstrate experiments that educate the students about the ways to reduce emissions. Students have sharp brains and by doing experiments, they might come up with innovative ideas to reduce greenhouse gases.

Climate change is not the kind of subject that only science students should pay attention to. Climate change affects the economy, health, and environment of the country. So, we all have to participate equally in this. We need to have an action plan to curb deforestation, floods, droughts, wildfires, and many more infectious diseases.

WHAT IS YOUR ACTION PLAN?

HIGHER EDUCATION INSTITUTES

Higher Education Institutes can play a vital role. Institutes provide funding for students to do various research in this field.

"Brilliant and new and true," said Rosanne Cash.

If students can research in their respective field, they are more likely to solve questions like: how to strategize in order to capture carbon and contribute to decarbonization? We need to understand that the investments we make today in research will pay us back in the future. We can do partnerships with the centres of health and the Global Environment Departments. "Where there is a will, there is a way." We have to endeavour to preserve our generations from upcoming pandemics.

We need dedicated people for research who are willing to do analysis, programming, and understand local culture and finances. Every single citizen is responsible for climate change, and it is vital to take action now.

PARTNERSHIP WITH ORGANISATIONS

Partnerships with private, government, and health sectors can accelerate the movement of awareness in society. The USA and European countries have been doing a lot of experimentation, examination, and exploration on this topic for many years. China has already implemented research that is improving everyday life because they know this one-time investment has long-term advantages. Energy efficiency not only saves your environment but your economic survivability. All the organisations should work in favour of public health in order to create a society free from climate related pandemics. Universally, we should aim to make a carbon-neutral society by 2050.

BANKS

Banks can aid us financially. The World Bank is already supporting agriculture by providing investments in irrigation infrastructure. With increasing temperatures and population, we require more water. Crop irrigation is important to provide food security for the population. Over time the World Bank will work towards the efficient usage of water and try to abolish hunger.

NON - GOVERNMENTAL ORGANIZATION

The government may not be able to provide efficient funding for such research as it already has various responsibilities like defence. In such cases, NGOs can participate actively and create awareness about climate change.

In general, we need support to be dedicated to the climate change disaster. Additionally, people need to make some changes in their own day-to-day lives. During COVID -19, people changed their lifestyle to stay safe, then why not now?

"It's easy to feel powerless in the face of a problem as big as climate change. But you're not powerless. And you don't have to be a politician or a philanthropist to make a difference." said Bill Gates.

I have written this book because I feel a lot of us are still unaware of this upcoming pandemic. I want you to be accountable for your actions, to understand how powerful you are. Try to use your power, mind, and strength for the betterment of this world.

WHAT NOW?

1. GO ONLINE AND POST HOW YOU ARE GOING TO REDUCE CARBON EMISSIONS WITH THE TAG #REDUCECARBONEMISSIONS.

2. PLAN AN EVENT #REDUCECARBONEMISSIONS WITH FAMILY OR FRIENDS. USE THIS BOOK AS A GUIDE AND SHARE YOUR IDEAS, YOUR CURIOSITIES AND TEACH ONE ANOTHER.

3. GIVE A COPY OF THIS BOOK AWAY TO SOMEBODY WHO NEEDS TO READ IT.

BIBLIOGRAPHY

By 2100, Deadly heat may threaten the majority of Humankind. Up to 75 percent of people could face deadly heat waves by 2100 unless carbon emissions plummet, a new study warns.
(2017)https://www.nationalgeographic.com/science/article/heat waves-climate-change-global-warming

Green office environments are linked with higher cognitive function scores.;2015
https://www.hsph.harvard.edu/news/press-releases/green-office-environments-linked-with-higher-cognitive-function-scores/

A changing climate worsens Allergy symptoms, 2010
https://www.ucsusa.org/resources/climate-change-worsens-allergy-symptoms

UBCO researcher uses geology to help astronomers find habitable planets, 2021
https://news.ok.ubc.ca/2021/05/04/ubco-researcher-uses-geology-to-help-astronomers-find-habitable-planets/

Climate change impacts on planktonic foraminifera through time 2021
https://www.Earth.ox.ac.uk/2021/04/climate-change-impacts-on-planktonic-foraminifera-through-time/

Coronavirus, climate change, and the environment A conversation on covid-19 with dr.Aaron Bernstein, director of Harvard chan C-Change
https://www.hsph.harvard.edu/c-change/subtopics/coronavirus-and-climate-change/

India ranks third in the global obesity index
https://www.einnews.com/pr_news/486386188/india-ranks-3rd-in-global-obesity-index

How does Ozone damages Plants?
https://scied.ucar.edu/learning-zone/air-quality/how-does-ozone-damage-plants

The Paris Agreement
https://unfccc.int/process-and-meetings/the-paris-agreement/the-paris-agreement

Water in agriculture
https://www.worldbank.org/en/topic/water-in-agriculture

Air pollution,
WHO.https://www.who.int/health-topics/air-pollution#tab=tab_3

Ebola virus
https://www.who.int/health-topics/ebola/#tab=tab_1

The climate crisis- covid 19: A major threat to the pandemic response
https://www.hsph.harvard.edu/c-change/news/the-climate-crisis-and-covid-19-a-major-threat-to-the-pandemic-response

Global burden of air pollution
http://www.healthdata.org/sites/default/files/files/infograp
hics/Infographic_AAAS_Air-pollution_2016.pdf

As carbon dioxide levels climb, millions are at risk of
nutritional deficiency.
https://www.hsph.harvard.edu/news/press-releases/climate-
change-less-nutritious-food/

Cutting red meat- for longer life.
https://www.health.harvard.edu/staying-healthy/cutting-red-
meat-for-a-longer-life

Temperature related death and illness
https://health2016.globalchange.gov/temperature-related-de
ath-and-illness

A field experiment with elevated atmospheric carbon dioxide
changes to C4 crop- herbivore interactions
https://www.nature.com/articles/srep13923

Climate change and health
https://www.who.int/news-room/fact-sheets/detail/climate-
change-and-health

How scientists estimate 'climate sensitivities.
https://www.carbonbrief.org/explainer-how-scientists-estima
te-climate-sensitivity

https://pixabay.com/?utm_source=link-attribution&ut
m_medium=referral&utm_campaign=image&utm
_content=1093758

https://pixabay.com/users/klimkin-1298145/?utm_source=link-attribution&utm_medium=referral&utm_campaign=image&utm_content=1093758

https://pixabay.com/users/goranh-3989449/?utm_source=link-attribution&utm_medium=referral&utm_campaign=image&utm_content=4057783

https://pixabay.com/users/dmncwndrlch-3075312/?utm_source=link-attribution&utm_medium=referral&utm_campaign=image&utm_content=4891275

https://pixabay.com/?utm_source=link-attribution&utm_medium=referral&utm_campaign=image&utm_content=1182713

https://pixabay.com/?utm_source=link-attribution&utm_medium=referral&utm_campaign=image&utm_content=4061206

https://pixabay.com/users/sasint-3639875/?utm_source=link-attribution&utm_medium=referral&utm_campaign=image&utm_content=1782430

https://pixabay.com/users/stocksnap-894430/?utm_source=link-attribution&utm_medium=referral&utm_campaign=image&utm_content=926308

https://pixabay.com/users/cris_f-9973234/?utm_source=link-attribution&utm_medium=referral&utm_campaign=image&utm_content=4387116

https://pixabay.com/users/pexels-2286921/?utm_source=link-attribution&utm_medium=referral&utm_campaign=image&utm_content=2178586

https://pixabay.com/users/castagnari53-15149183/?utm_source=link-attribution&utm_medium=referral&utm_campaign=image&utm_content=4845841

https://pixabay.com/users/caniceus-15612619/?utm_source=link-attribution&utm_medium=referral&utm_campaign=image&utm_content=5663191

https://pixabay.com/users/jniittymaa0-701650/?utm_source=link-attribution&utm_medium=referral&utm_campaign=image&utm_content=1185076

https://pixabay.com/users/pexels-2286921/?utm_source=link-attribution&utm_medium=referral&utm_campaign=image&utm_content=1838658

https://pixabay.com/?utm_source=link-attribution&utm_medium=referral&utm_campaign=image&utm_content=2367029

https://pixabay.com/users/comfreak-51581/?utm_source=link-attribution&utm_medium=referral&utm_campaign=image&utm_content=1398064

https://pixabay.com/?utm_source=link-attribution&utm_medium=referral&utm_campaign=image&utm_content=3313646

https://pixabay.com/?utm_source=link-attribution&utm_medium=referral&utm_campaign=image&utm_content=4051083

https://pixabay.com/users/dmncwndrlch-3075312/?utm_source=link-attribution&utm_medium=referral&utm_campaign=image&utm_content=4891275

https://pixabay.com/users/markmags-2013644/?utm_source=link-attribution&utm_medium=referral&utm_campaign=image&utm_content=1182713

https://pixabay.com/users/sasint-3639875/?utm_source=link-attribution&utm_medium=referral&utm_campaign=image&utm_content=1782430

https://pixabay.com/users/stocksnap-894430/?utm_source=link-attribution&utm_medium=referral&utm_campaign=image&utm_content=926308

https://pixabay.com/users/pexels-2286921/?utm_source=link-attribution&utm_medium=referral&utm_campaign=image&utm_content=2178586

https://pixabay.com/?utm_source=link-attribution&utm_medium=referral&utm_campaign=image&utm_content=872000

Image by <a href="https://pixabay.com/users/vintagelee-15400574/?utm_source=link-attribution&utm_medium=referral&utm_campaign=image&utm_content=5832393">Robert

Lee Cortes</a> from <a
href="https://pixabay.com/?utm_source=link-attribution&a
mp;utm_medium=referral&utm_campaign=image&am
p;utm_content=5832393">Pixabay</a>
  Image by <a
href="https://pixabay.com/users/mollyroselee-9214707/?ut
m_source=link-attribution&utm_medium=referral&am
p;utm_campaign=image&utm_content=4469698">moll
yroselee</a> from <a
href="https://pixabay.com/?utm_source=link-attribution&a
mp;utm_medium=referral&utm_campaign=image&am
p;utm_content=4469698">Pixabay</a>

Man and pollution - Photo by João Cabral from Pexels   one
world - Photo by Markus Spiske from Pexels   drought -
Photo by Laker from Pexels

Lost in other planets - Photo by Tom Leishman from Pexels
pollution Photo by Anna Shvets from Pexels   Photo by Tom
Leishman from Pexels Photo by Markus Spiske from Pexels

Photo by Markus Spiske from Pexels Photo by Guillaume
Falco from Pexels Photo by Markus Spiske from Pexels
Photo by Markus Spiske from Pexels Photo by Markus
Spiske from Pexels Photo by Mikhail Nilov from Pexels
Photo by Ron Lach from Pexels Photo by Amir Esrafili from
Pexels Photo by energepic.com from Pexels

https://pixabay.com/users/vinsky2002-1151065/?utm_sour
ce=link-attribution&utm_medium=referral&utm_c
ampaign=image&utm_content=5032718

https://pixabay.com/users/arttower-5337/?utm_source=link-attribution&utm_medium=referral&utm_campaign=image&utm_content=5288667

https://pixabay.com/users/darkmoon_art-1664300/?utm_source=link-attribution&utm_medium=referral&utm_campaign=image&utm_content=6544798

https://pixabay.com/users/gdj-1086657/?utm_source=link-attribution&utm_medium=referral&utm_campaign=image&utm_content=1254501

https://pixabay.com/users/shrikeshmaster-4115921/?utm_source=link-attribution&utm_medium=referral&utm_campaign=image&utm_content=4995808

https://pixabay.com/users/matryx-15948447/?utm_source=link-attribution&utm_medium=referral&utm_campaign=image&utm_content=5048097

https://pixabay.com/users/clker-free-vector-images-3736/?utm_source=link-attribution&utm_medium=referral&utm_campaign=image&utm_content=26622

https://pixabay.com/users/blende12-201217/?utm_source=link-attribution&utm_medium=referral&utm_campaign=image&utm_content=1480779

https://pixabay.com/users/squarefrog-9690118/?utm_source=link-attribution&utm_medium=referral&utm_campaign=image&utm_content=6058078

https://pixabay.com/?utm_source=link-attribution&ut
m_medium=referral&utm_campaign=image&utm
_content=2208079

https://pixabay.com/illustrations/search/?pagi=31&cat=sci
ence&orientation=vertical&

https://pixabay.com/users/mysticsartdesign-322497/?utm_s
ource=link-attribution&utm_medium=referral&ut
m_campaign=image&utm_content=761821

ABOUT THE AUTHOR

NANDITA KHANDELWAL is a writer who loves to dance. She is a Geologist, Health Coach, and more. She lives in New Delhi, India, and is online on Instagram and YouTube Channel @Nanditakhandelwal.

NOTES